The book no one can read

By Joni Morton

After 500 years this book to this day remains unread...

The manuscripts cover front and back (scene left to right)

Finding an old manuscript in a chest, in a library, in a medieval castle in Italy is strange to say the least. And that is just the beginning.

For over 100 years, experts and amateur researchers have tried to solve the riddle of a handwritten book, referred to as the book no one can read or the Voynich manuscript. It is composed in an unknown script, and is handwritten, consisting of 246 pages containing numerous illustrations and approximately 170,000 characters. What is unique about it, is that the script is utterly unknown and therefore illegible.

The late medieval manuscript is written in a cramped no punctuation script and illustrated with lively line drawings that have been painted over, at times crudely, with washes of color. These illustrations range from heavy-headed flowers that bear no relation to any earthly variety to the bizarre naked and possibly pregnant women, frolicking in what look like amusement-park water slides from the fifteenth century. With their distended bellies, stick-like arms and legs, and earnest expressions, the naked figures have a whimsical quality, though their anatomy is frankly

something unusual for the period. The manuscript's botanical drawings are no less strange, the plants appear to be chimerical, combining incompatible parts from different species, even different kingdoms. Tentacled balls of roots take the forms of animals, or of human organs—in one case, sprouting two disembodied heads with vexed expressions.

Approximately 220 of the 246 pages are illustrated. Some of the pages can be unfolded, revealing illustrations that extend to several page lengths.

Because, unlike the text, the illustrations can be divided into different sections. Six chapters of the manuscript can be distinguished: the botanical chapter with large plant illustrations, the astronomical chapter with charts containing celestial bodies and the zodiac signs, the balneological chapter (the science of baths or bathing) with nude female figures in tubs, the cosmological chapter with circles and rosettes, the pharmaceutical chapter with plants, parts of plants, and pots, as well as a chapter with food recipes.

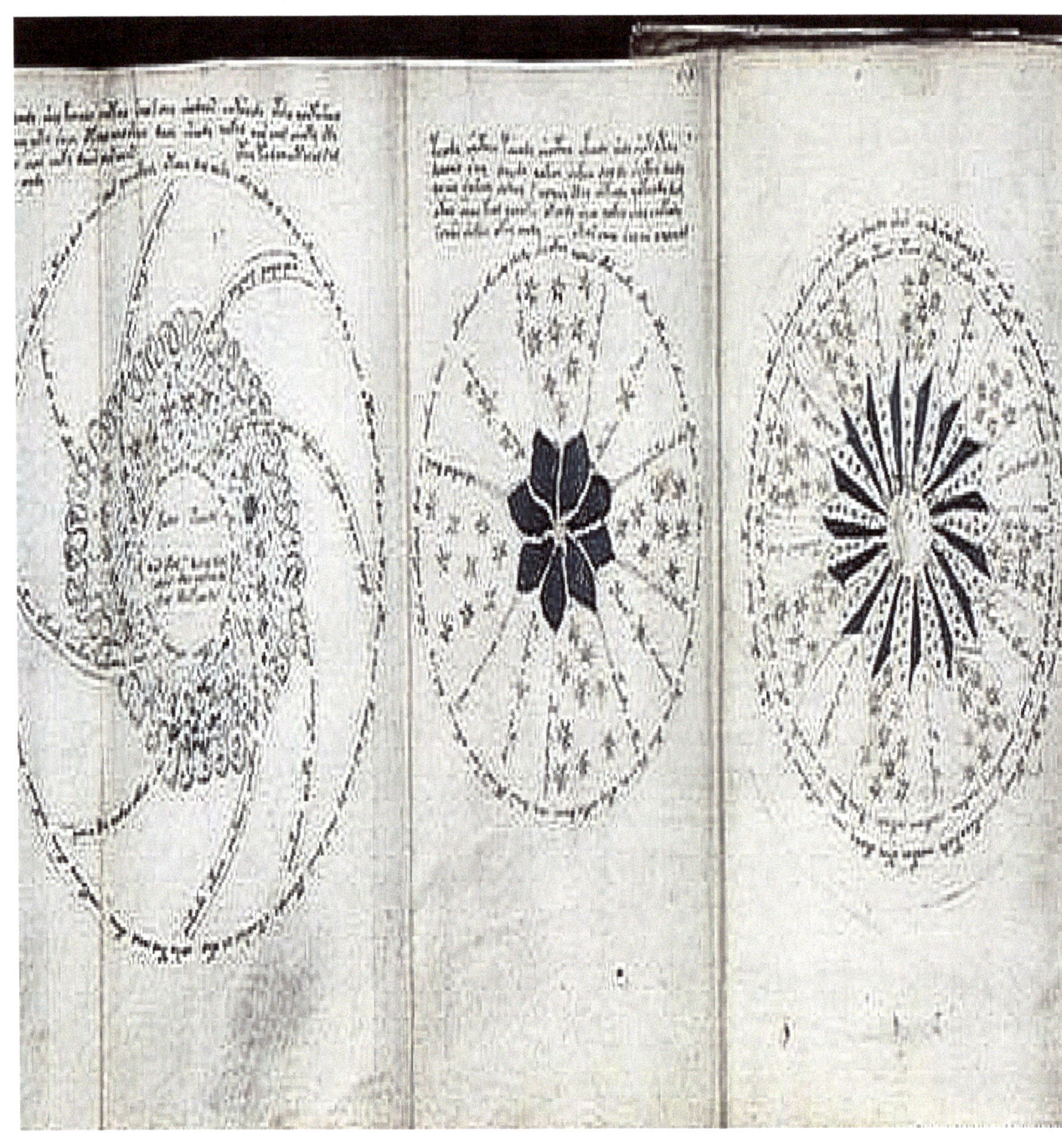

The celestial bodies illustrated in the astronomical section cannot be identified.

The hairstyles and clothing of the people pictured in the book, as well as the style of the illustrations, were usually dated to the period 1450–1520, which proved reasonably compatible with the radiocarbon dating (between 1404 and 1438).

There are between fifteen and twenty-five different letters in the manuscript, but in many cases it is not clear whether identical or different symbols have been used.

For this same reason, letter frequency cannot be determined clearly.

Nevertheless,

the language of the manuscript can be brought in line with European languages, because the average word length is four or five letters. Following this line of consideration, arguments can be put forward that Greek, Latin, or one of several other European languages was used to compose the manuscript.

It is a pity that this approach does not implicate a specific language.

It is curious that some words are repeated successively up to five times.

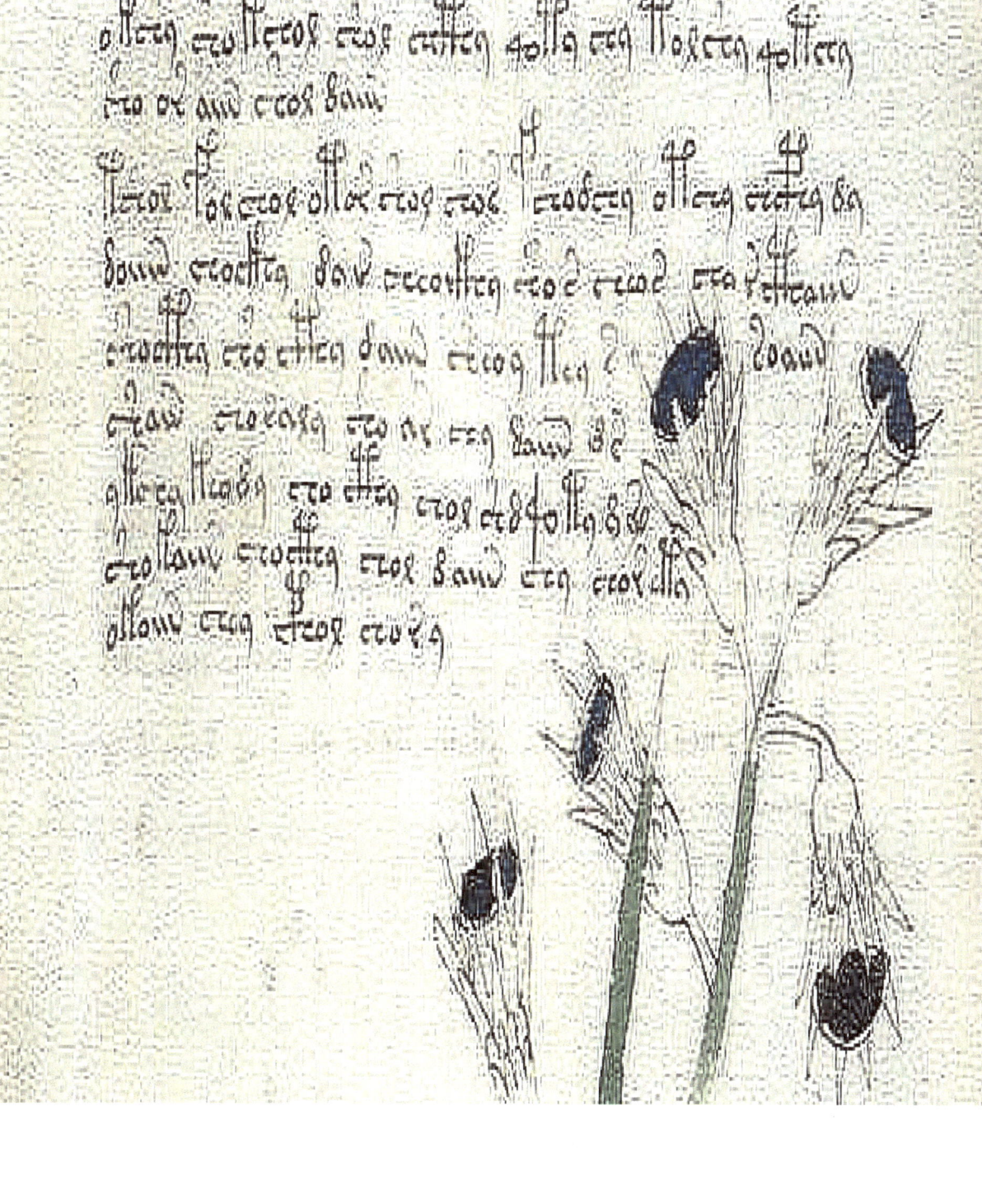

The distribution of the letters within each word also does not answer known language patterns. Looking at the text as a whole, far fewer recurring words turn up than would be expected. Such arguments reveal with a high probability that against all appearance to the contrary we are not dealing with a simple substitution of letters. There also is no clear evidence that other simple encryption methods were used.

No one knows who wrote the book, what it contains, and what its purpose was. The author of the manuscript wrote from left to right, this can be discerned from the left-aligned formatting. The typeface and size of the characters are inconspicuous, which is not altered by the fact that the text contains no punctuation marks, because this is unexceptional for old texts.

Thus it is evident to a layman, even before inspection of the illustrations, that the manuscript has its origins in European culture. Also, it is apparent that the author was very accurate. There are no visible corrections in the text.

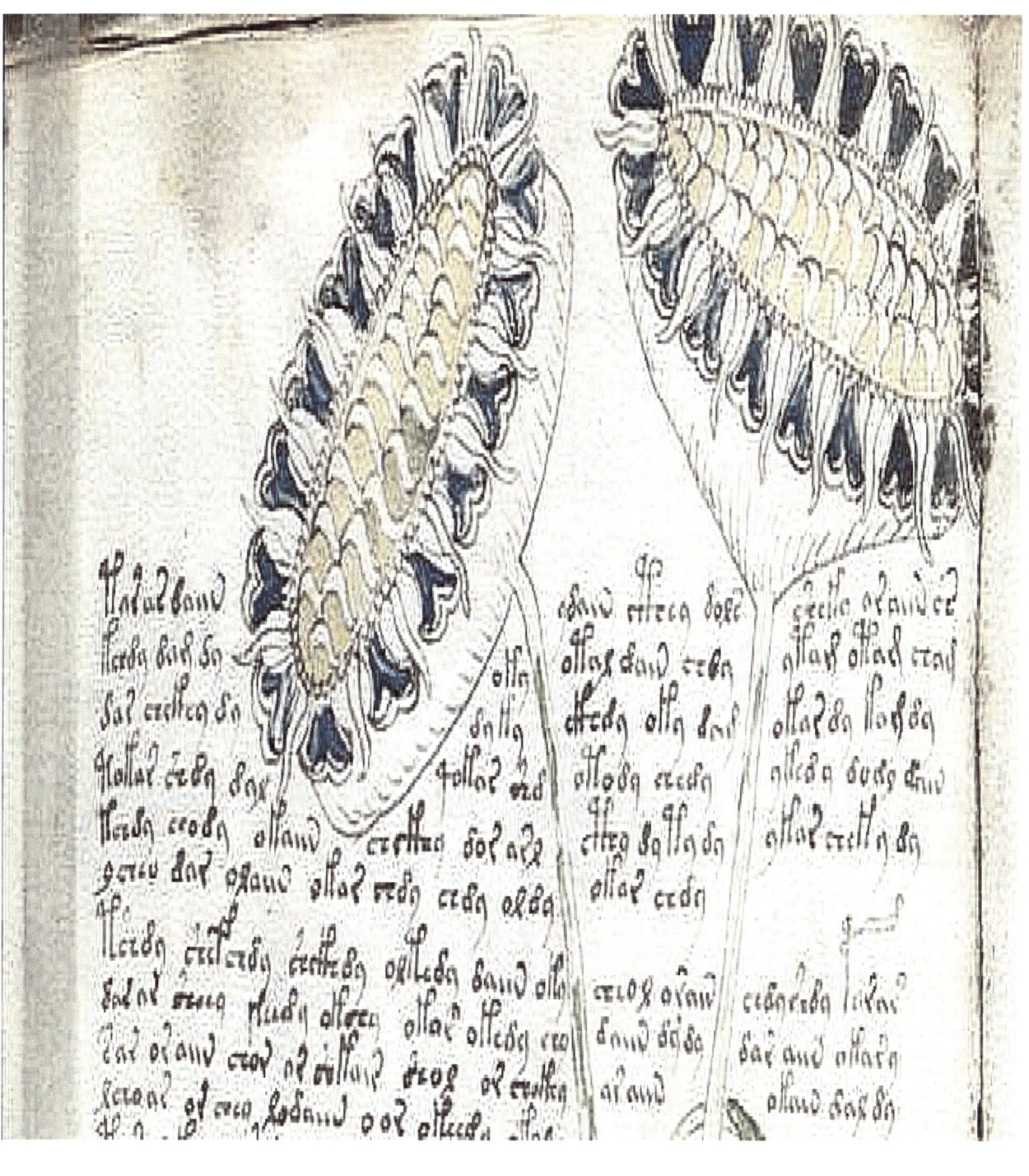

The text itself is not divided into chapters and there are no subheadings.

The different illustrations can hardly be related to a common topic. Therefore the manuscript, if it has meaningful content at all, it must be a treatise on many different subjects. One may possibly say that it is a textbook for magicians, physicians, pharmacists, and astrologers – these professions were not common 500 years ago. Provided that it is hardly possible to recognize significant symbols and religious motives within it, the manuscript can neither be assigned to a certain school of thought nor to a particular religion.

Unfortunately, none of the 126 plant illustrations can be definitively identified. However, the plant pictures at least enabled certain conclusions regarding the date of origin, before the radiocarbon dating was performed. Comparisons of artistic styles showed that the manuscript presumably did not originate before the fourteenth century, which was later confirmed.

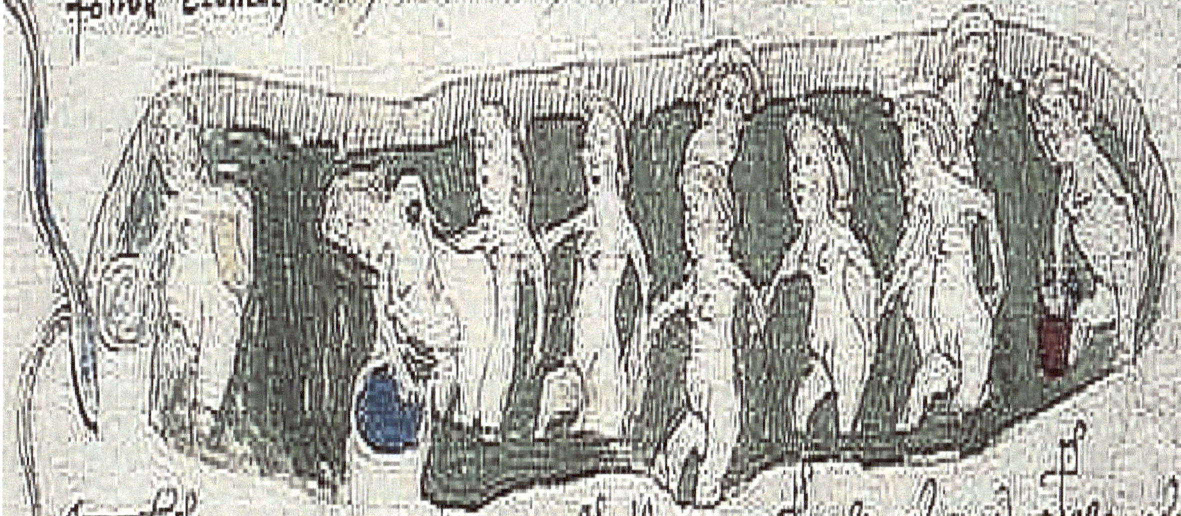

Now a little about the man that found the manuscript.

Wilfrid Voynich, born November 12, 1865 was an ethnic Pole from the Russian Empire, he was imprisoned in Siberia for revolutionary activities. He escaped the prison walls and made his way to London via Manchuria and China. This is where he met his future wife, Ethel Boole. Ethel Lilian Voynich, née Boole was an Anglo-Irish novelist and musician, and a supporter of several revolutionary causes. She was born in Cork, Ireland but grew up in England. Ethel became a popular novelist, and the couple set up a bookshop that was rumored to be a secret front for an organization against the tsarists regime.

Voynich, wanting to add rare books to the bookshop, in 1912 went on a buying trip to the Villa Mondragone. The villa is a patrician villa originally in the territory of the Italian commune of Frascati in central Italy. Now in the territory of Alban Hills.

Polish book dealer Wilfred M. Voynich.

Part of the art and antiquities collections there including the Antinous Mondragone which derives its name from the villa.

The facility, in need of funds, was discreetly selling some of its holdings.

Looking through old books in the castles library Voynich saw an old dusty chest and after inpsection, his eyes laid upon an old leather bound manuscript. With the manuscript he bought a total of 30 books, and brought them back to America.

We don't know is how to read the book or what it means.

What we do know is that there is a letter from 1666 that was found inside the cover that states that the manuscript had once belonged to the Holy Roman Emperor Rudolf II (1552-1612) and that he paid 600 gold ducats to buy it, around ninety thousand dollars in today's money. The book was already nearly two centuries old at the time of his purchase.

This letter found in the manuscripts cover is shown (below) it is known as the Marci Letter.

Reverend Lord Father in Christ

{This book was relinquished to me by a singular friend in his will, soon after my possession began I destined the same to you my dear Athanasius: indeed I am persuaded none but you can read it. The then possessor has sent you letters seeking your judgment about it in which he transmitted his description of it, for he was convinced it could be read by you; the true book in which he put such untiring work into deciphering he objected to sending, attempting to use the same experience you are sent here not giving up hope until his life reached its final limit. But the effort was in vain, for such sphinxes obey only their own. Please accept what was long due to you now as some small token of my affection for you, and may you break through its bars with your habitual ease. Doctor Raphael, the Czech language tutor of King Ferdinand III as they both then were, once told me that the said book belonged to Emperor

Rudolph and that he presented 600 ducats to the messenger who brought him the book. He, Raphael, thought that the author was Roger Bacon the Englishman. I suspend my judgment on the matter. You be the judge of what we should think about it. I commend myself to your favor.}

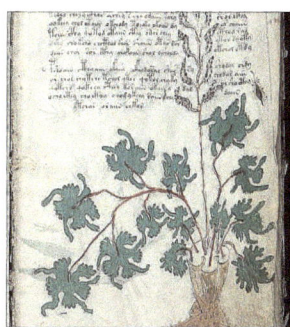
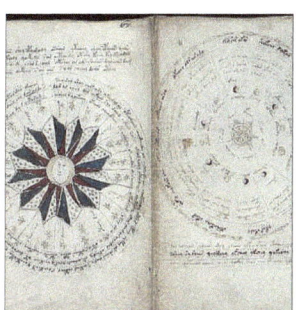
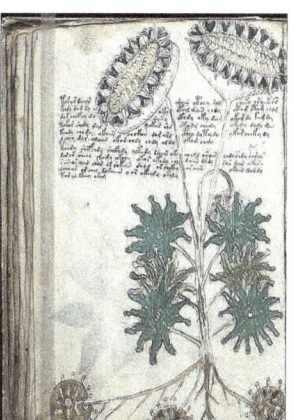

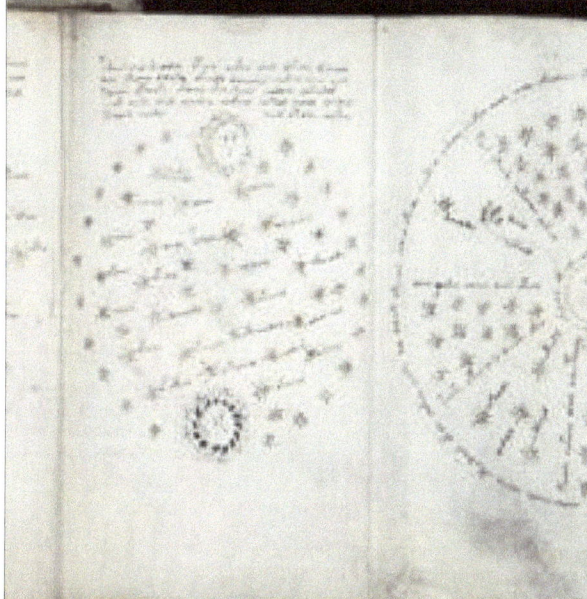
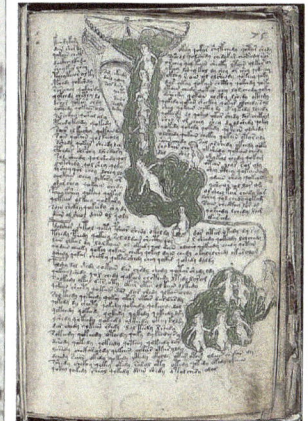

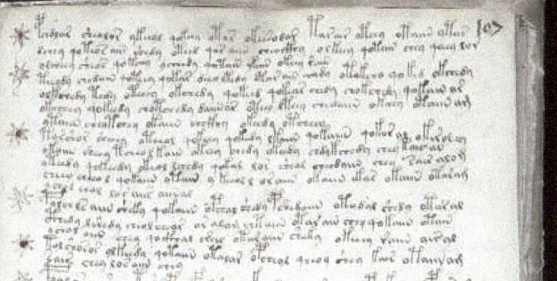
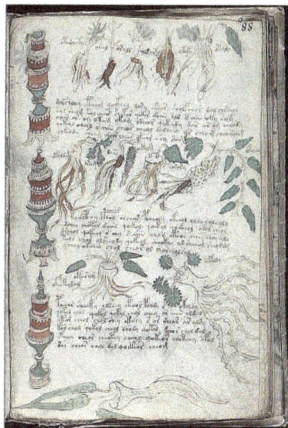

It's a near certainty that the Voynich manuscript, if ever translated, will yield great secrets and lost knowledge. But the fact that it really is as old as presumed means that it has historical value, if this mystery is ever unraveled it will at the very least tell modern researchers and historians something of the author, that created a book so baffling that it has not yielded to prying eyes for over 500 years.

Voynich spent the rest of his life trying to decipher are to find anyone that could decipher the "book that no one can read."

After Voynich's death in 1930 – New York, his widow inherited the book, and in 1969 Hans P. Kraus bought it and donated it to Yale University, which still owns it today.

It is cataloged under call number MS 408

A word from me, I really hope you enjoyed reading about this mysterious book, and take a minute of your time to make a quick review. When you leave a review it helps to push my books up into the popular books list, where more readers find out about them... it really helps new authors compete with the mega-conglomerate publishing companies, allowing folks like me to follow a dream.

I you loved or at least liked the book, thanks for taking a chance.

I sincerely appreciate it!

Please enjoy some of my other work found on – amazon.com/author/jonimortonJoni Morton – lives in Cleveland, Texas and is the author of Children's Books, Myths and Legends, DIY, and Health and Beauty Books. She is published on E-Books and in print through Amazon Publishing.

♥

Text copyright © Joni Morton 2016

All rights reserved. Any unauthorized reprint or use of this material is prohibited. No part of this book may be reproduced or transmitted in any form or by any means, electronic or mechanical, including photocopying, recording, or by any information storage and retrieval system without express written permission from the author/publisher. Your understanding is greatly appreciated.

www.ingramcontent.com/pod-product-compliance
Lightning Source LLC
Chambersburg PA
CBHW050439180526
45159CB00006B/2596